Water

73

空中的云朵

Clouds in the Sky

Gunter Pauli

[比] 冈特·鲍利　著

[哥伦] 凯瑟琳娜·巴赫　绘

郭光普　译

上海远东出版社

丛书编委会

主　　任：田成川

副主任：何家振　闫世东　林　玉

委　　员：李原原　翟致信　靳增江　史国鹏　梁雅丽
　　　　　任泽林　陈　卫　薛　梅　王　岢　郑循如
　　　　　彭　勇　王梦雨

特别感谢以下热心人士对童书工作的支持：

匡志强　宋小华　解　东　厉　云　李　婧　庞英元
李　阳　刘　丹　冯家宝　熊彩虹　罗淑怡　旷　婉
杨　荣　刘学振　何圣霖　廖清州　谭燕宁　王　征
李　杰　韦小宏　欧　亮　陈强林　陈　果　寿颖慧
罗　佳　傅　俊　白永喆　戴　虹

目录

Contents

ZERI Learning Initiative

小老鼠和猫头鹰

又见面了。猫头鹰喜欢吃老鼠，但不吃这一只，因为他喜欢和这只小老鼠聊天。小老鼠虽然不喜欢科学和数学，却问了很多聪明的问题。这些问题让猫头鹰思考。

"嗨！你好！"猫头鹰说，"你这几天好吗？你为什么看着天空发呆？"

The rat and the owl meet again. The owl likes to eat rats, but not this one, as he likes to talk to him. Although the rat does not like science and mathematics he asks a lot of smart questions. Questions that make the owl think.

"Oh, hello," says the owl. "How are you and why are you looking at the sky?"

你为什么看着天空发呆?

Why are you looking at the sky?

由亿万个很小很小的水滴构成

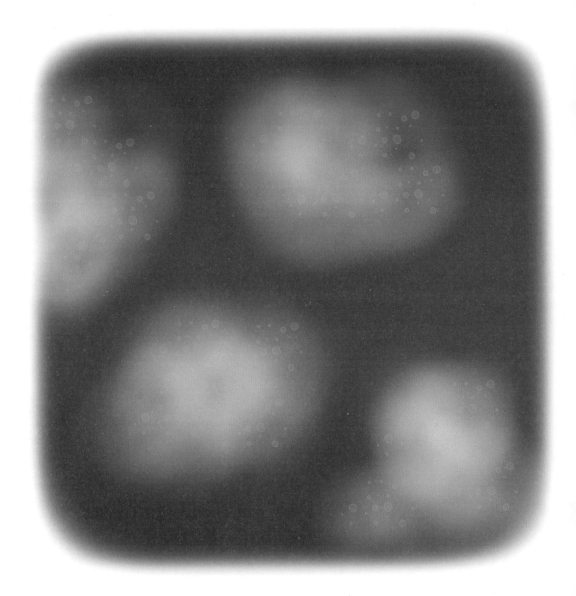

Made up of millions of tiny water drops

"我对云着迷了。"小老鼠回答道，"云的存在使万有引力定律看上去失效了。"

"云是由天空中亿万个很小很小的水滴构成的。"

"我们知道苹果不会飞，它们成熟后就会掉到地上，但我还不知道水居然会飞！"

"I'm fascinated by the clouds," responds the rat. "Clouds show that the law of gravity does not really work."

"Clouds are made of millions of tiny water drops up there in the sky."

"We know apples don't fly, they just drop to the ground when ripe, but I didn't know water could fly!"

"水当然不会飞，但是云只要比周围的空气温度高就会漂浮在空中。"

"但你刚才说云是由水做的，水可比空气重呀！"

"但你看，那些水滴太小了，它们遇到的空气阻力很大，减慢了它们的下降速度。"

"Water doesn't fly, but clouds will float as long as they are warmer than the air around them," Owl explains.

"But you just said clouds are made of water. That's heavier than air."

"But you see, those droplets are so small that they experience a lot of air resistance and that slows their fall."

比周围的空气温度高

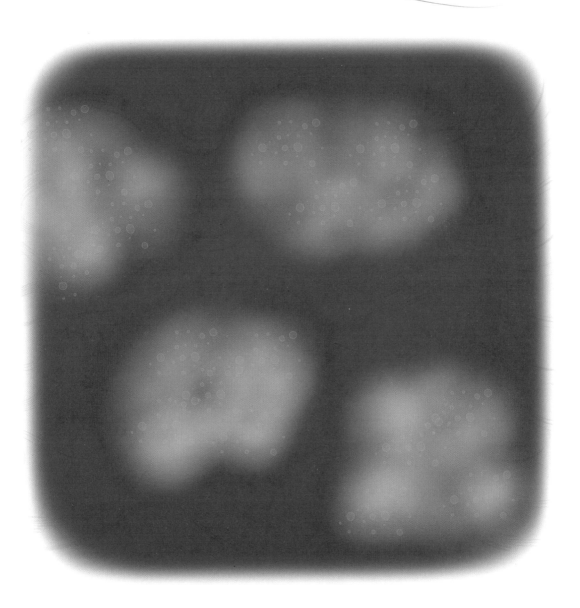

Warmer than the air around them

因为还有风

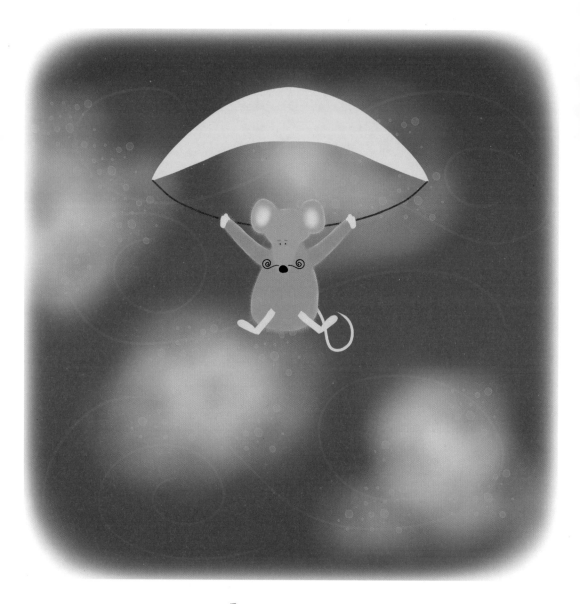

Because there is also the wind

"那你怎么解释

云漂浮在空中，很少掉下

来呢？"

"因为还有风呀。"猫头鹰说道。

"天上到处都有水蒸气，为什么从这

里到那里只有一片云，而不是很多

云到处漂浮呢？"

"So how do you explain
clouds floating in the sky and
seldom dropping to the ground?"
"Because there is also the wind,"
says Owl.
"And when there's evaporated water
everywhere in the sky, why does
only one cloud form here or
there, and not lots of clouds
everywhere?"

"因为小水滴需要
粘在小灰尘上才能形成云。"

"我明白了。"小老鼠说，"所以云
是水和灰尘构成的。那你告诉我，为什么
有些云是白色的，还有些是灰色的呢？"

"因为当太阳光照透云层时，云看起来就
是白色的。但是当云层太厚时，穿透
的太阳光太少了，云看起来就
是灰色的了。"

"Because water
droplets need to stick to tiny
dust particles to form a cloud."

"I see," says Rat. "So a cloud is
water and dust. Now tell me why are
some clouds white and others grey?"

"Because when light shines through
clouds, they look white. But when
clouds get too thick, less light
shines through and they look
grey."

云是水和灰尘构成的

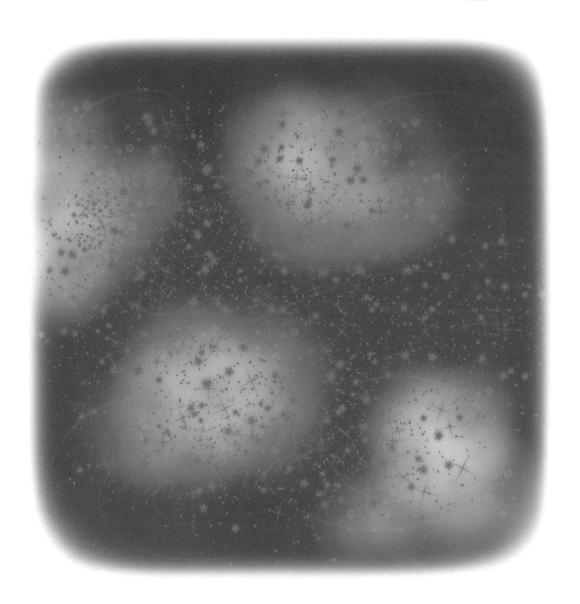

A cloud is water and dust

水是一个简单的分子

Water is a simple molecule

"哎呀! 猫头鹰先生, 你对云可真了解呀! "

"好吧，其实我真的只知道一点。另一方面，人们觉得自己对水很了解，但我却觉得他们一点儿也不懂水。"

"有人告诉我，水是一个简单的分子，是由2个氢原子和1个氧原子构成的。"

"My word, Mr Owl. You do know a lot about the clouds!"

"Well, I really only know a little. People, on the other hand, think they know a lot about water but I don't think that it is true."

"I've been told water is a simple molecule, made up of two hydrogen atoms and one oxygen atom."

"噢，是的，我们
完全了解什么是分子，但我们需要
进一步了解的是这么多聚在一起的分子是
如何相互作用的。只看分子，就像只看一根头
发却不看发型，而发型却能反映一个人的性格。"
"太对了。很多水分子紧紧挤压在一起时会像石
头一样坚硬，我们称它为冰。而当水以液体形
式在河流里流动时，我们就可以喝了。但
是水还能以气体形式存在，我们称之
为水蒸气。"

"Oh yes, of
course, we know all about
molecules, but we need to know
more about how a collection of many
of these molecules interact. Only looking
at the molecule is like looking at only
one hair on your head without seeing the
hairstyle that reflects your character."
"Too true. Lots of water molecules packed
closely together can be as solid as a
rock – we call that ice. And water flows
through rivers as a liquid – that we
drink. But water can also be in
the form of a gas – we call
that vapour."

这你都知道，太让我吃惊了

It amazes me that you know this

"这你都知道，太让我吃惊了。"猫头鹰说。

"但云到底是什么呢？是由很多小冰粒、小水珠和水蒸气组成的吗？"

"嗯，要是磁铁的两个正极靠近会发生什么事呢？"

"这个我知道！它们会互相排斥；两个负极也会相互排斥。"

"It amazes me that you know this," says the owl.

"But what is a cloud really? Are clouds made of lots of tiny balls of ice, drops of water, and also vapour?"

"Well, what happens when two positive poles of a magnet come close to each other?"

"I know! They repel each other. And negatives repel each other as well."

"太对了！但云
是靠什么把这些带负电的水分
子聚集在一起的呢？"

小老鼠沉默了一会儿，看着猫头鹰说
道："我觉得我得再观察云一段时间才能搞
清楚这个问题。"

"好极了！就让美丽的天空激励你去进
行新的科学探索吧！"

……这仅仅是开始！……

"Exactly,
but what makes the
clouds with all these negative
water molecules hold themselves
together?"

After a moment of silence, the rat looks
at the owl and says, "I think I will have to
look at the clouds a little bit longer to figure
that one out."

"Great. Let the beauty of the sky
inspire you to make new scientific
discoveries!"

... AND IT HAS ONLY JUST
BEGIN!...

······这仅仅是开始！······

...AND IT HAS ONLY JUST BEGUN! ...

Did You Know?

你知道吗?

About 70% of your body weight is water. However, as water molecules are the lightest molecules in our body, approximately 99% of our body's molecules are water.

我们身体重量的70%都是水。然而,因为水分子是我们身体中最轻的分子,所以其实我们身体中将近 99% 的分子都是水分子。

A water molecule (H_2O) has two hydrogen atoms and one oxygen atom. However, when water molecules cluster together it also behaves like $H_3O_2^-$.

一个水分子(H_2O)包含两个氢原子和一个氧原子。但是,当水分子聚集在一起的时候,也能形成 $H_3O_2^-$。

Fog is a type of cloud. It is created when warm, moist air flows over colder soil. If the air is saturated, the moisture condenses and forms fog that could reduce visibility to near zero.

雾是云的一种形式。它是由温暖潮湿的空气流过温度较低的地面形成的。如果空气中的水蒸气饱和了，水分就会凝结并形成雾，并使能见度几乎降为零。

There are 10 types of clouds and some can move at more than 100 km/hr.

云有 10 种，其中一些云的移动速度能达到 100 千米／时。

When air is heated by the sun, it rises and slowly cools down, and when it reaches saturation point it condenses, forming a cloud. The forces that create a cloud are pressure and temperature. Clouds are held together as a result of charged molecules.

当空气被太阳加热时，就会上升并慢慢冷却，当水蒸气饱和时就会凝结形成云。云的形成取决于压强和温度。云聚在一起是由于带电分子的作用。

While two negative or positive charges repel each other, negative charges accumulated at the bottom of the cloud.

两个负电荷或正电荷互相排斥，但云中的负电荷却聚集在云层下端。

Scientists argue that clouds, water bridges, and insects running on water can be explained through an understanding of the "fourth phase" of water.

科学家认为云、水桥和昆虫在水上行走可以用对水的"第四相"的理解来进行解释。

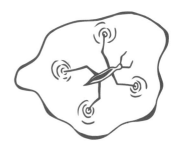

Water is one of the simplest molecules and also the most abundant molecules on Earth. Yet water is the least understood of all substances.

水是最简单的分子之一，也是地球上含量最多的分子；但它也是所有物质中人们最缺乏了解的。

Think about It

想一想

W ater expands as it gets warmer and contracts as it gets colder, but if it gets colder than 4 °C, it starts expanding again. Is this the rule or the exception?

水在变得温暖时就会膨胀，而在变冷时就会收缩，但是当温度降到 4℃以下时又开始膨胀。这是规律还是例外呢？

The owl knows a lot and still claims he knows very little. Is he modest?

猫头鹰懂得很多却仍然说自己知道得很少，他是在谦虚吗？

如果云可以制造闪电，那么只靠压强差和温度差是不是也能制造闪电呢？

If clouds produce lightning, can lightning be produced through pressure and temperature differentials only?

Do you think new scientific discoveries can be made by looking at the beauty in nature?

你认为通过观察自然界的美可以获得新的科学发现吗？

Let us make a bridge with water.

Fill two cups with water. Put the cups on isolated padding.

Make the water inside the cups connect. Then send a small electric current through the cups. Slowly move the cups apart. What will be left behind is a bridge of floating water.

Water flows from the positively charged cup to the negatively charged cup.

You can make the water bridge up to 2 cm long. If the water gets hot, the bridge will collapse.

让我们用水造一座桥吧。在两个杯子中装满水，放到分开的衬垫上。让杯子里的水连接起来；再通一股微弱电流，慢慢把杯子移开，就会发现杯子间有一条漂浮的水桥。水将从连接正极的杯子流向连接负极的杯子。你可以把水桥做到 2 厘米长。如果水变热了，水桥就会断掉。

学科知识
Academic Knowledge

生物学	鸟如何利用暖气流而不是自己的能量飞得更高；撒哈拉沙漠的尘土如何来到亚马孙三角洲并成为表层土壤；生命离不开水，水是所有生物的主要成分。
化 学	H_2O和$H_3O_2^-$的区别；氢原子（H）和氧原子（O）。
物 理	云的作用；万有引力定律；悬浮在空中的水滴产生的摩擦力；灰尘颗粒和小水滴之间的吸引力；比较一个分子的行为和一堆分子的行为；热空气的膨胀；空气中水的饱和；云的色彩有白色和灰色；云的10种形态；原子和分子的不同之处；磁性物质同极相斥，异极相吸；比较压强和温度的结合同压强、温度和电荷的结合；水滴的大小决定其物理特性；风是怎样形成的；打雷和闪电；水的禁区——上文提到的水的"第四相"对人类健康和地球健康的核心作用；水的"第四相"的发现超越了物理学的范畴，进入了化学和生物学的范畴；水和光的相互作用。
工程学	天气和气候工程学能提高天气预测能力。
经济学	云计算是储存和处理数据和信息的新形式。
伦理学	要接受科学上新的视角，这将不断激励我们彻底改变我们周围的世界。
历 史	威廉·阿姆斯特朗在1850年第一次展示了水桥（也被称作水绳）。
地 理	利用气象学决定农事。
数 学	当更多的参数互相作用时，复杂性会增加；光和声的速度。
生活方式	俗话说："光打雷不下雨。"就是说只有承诺却没有实现。
社会学	在希腊神话中，仙女和气象之神生活在云里；在北欧神话中托尔是雷神，而在希腊神话中涅斐勒是云神；儒家学说认为通过不道德手段获得的财富和地位只是浮云。
心理学	发型可以表达个性；艺术和科学的协同作用：科学家已经发现音乐，尤其是莫扎特的音乐能够促进智力发展，如数学思考。
系统论	自然系统中的相互作用不能被简单化，就像云的形成不能仅仅用温度和压强来解释，还需要考虑灰尘颗粒和水滴的摩擦力，此外还有磁力。

情感智慧
Emotional Intelligence

猫头鹰

猫头鹰变得更谦虚了。开始的时候，猫头鹰认为自己是个聪明人，他让小老鼠活了下来。现在猫头鹰认识到小老鼠的存在并准备和他交流。他兴趣盎然地用一个问题开始了这次交谈。猫头鹰无私地和小老鼠分享了自己的知识，甚至很高兴回答小老鼠的所有问题。当小老鼠抛开细节看到整个问题时，猫头鹰很是吃惊。于是他走进了一个未知的领域，把云和磁联系起来，还引入了一些虽然已经研究了一个多世纪但对于大多数科学家还很新的概念。然后猫头鹰表示了对小老鼠的支持并鼓励他自由思考，去拥抱自然和艺术的美丽。

小老鼠

小老鼠表现得比较成熟，并且在猫头鹰面前很少挑衅。他没有挑战猫头鹰，而是从猫头鹰那里获得了知识，而且学得很快。小老鼠对猫头鹰的聪明才智表现出了敬佩和赞赏。当猫头鹰让小老鼠进一步思考的时候，小老鼠愉快地接受了挑战；当猫头鹰称赞小老鼠丰富的知识时，小老鼠并没有为此表示感谢，而是继续提出了更困难的问题。但他很快便用完了自己的知识。然而，他依然保持着自信，他抓住了移动的云朵给他的灵感，继续学习更多的知识。

艺术
The Arts

在一大张白纸上画出至少6种不同形状的云。你不用知道这些云的科学名称。在你相信会先落下雨的云朵下面画上水滴。再选一张你认为一定会产生闪电的，画上闪电。最后选一张，画上彩虹。

思维拓展
Systems: Making the Connections

世界是复杂的。为了更好地理解组成整个现实生活的很多小的部分，科学已经为我们把现实简化了。仅仅描述氢原子和氧原子的特点无法解释水分子的行为。很多水分子结合在一起时，在不同的压强和温度环境下会有不同的表现。然而，这还仅仅是影响物质行为的两个变量，我们知道还有其他因素（如磁力和电荷）也影响着所有物体，包括水和云，正如我们在打雷和闪电时看到的。但是，由于我们不知道如何测量云中的电荷，所以我们忽略了一些事实，尽管我们知道它们的存在。虽然一些现象在科学假说中得到了解释，但是我们知道现实世界要复杂得多。根据万有引力定律，自然的力量能使苹果从树上掉下来。但我们是否理解使苹果首先从树上长出来的所有力量是什么？我们将现实过度简单化以更好地理解自然现象，去理解水为什么在冰冻时膨胀，但同时我们也要面对另外一些无法解释的现象。还有很多现象我们无法理解，比如为什么很多水分蒸发到空中各处却只形成了一朵云？为什么水桥在水温上升时会断掉？为什么蜥蜴、蜘蛛和水黾能在水上行走？

动手能力
Capacity to Implement

对于那些用我们学过的简单科学知识无法解释的现象，我们必须仔细考虑现象的复杂性和它们之间的相互联系。这就是为什么我们要理解我们每天都能观察到的云。当我们大胆地对生活的复杂性和所有关系进行深入理解时，必须认识到今天的科学还不能解释一切。我们将有新的发现并将创立基于更多高深的假说上的各种新理论。这需要强烈的求知欲、创造力和灵感：在幻想、想象和现实之间探索。有时艺术家能够帮助我们进行探索并超越已经建立起来的科学的边界。他们就像社会的天线，能感觉到无法解释的事物，并提供找到答案的途径，一条不同寻常的途径。然后就由科学家进行深入研究，提出理论和证据，对我们周围的现实给出了新的理解。

故事灵感来自
This Fable Is Inspired by

杰拉尔德·波拉克
Gerald Pollack

　　杰拉尔德·波拉克博士本来在纽约大学电气工程专业学习，后来他转入宾夕法尼亚大学并获得了生物医学工程学位。他大胆地开始了有关水、细胞和健康的新的科学研究。现在他是华盛顿大学西雅图分校波拉克实验室的负责人。他研究运动、细胞生物学以及生物表面和水的相互作用。他的两部著作《肌肉和分子》及《胶体和生命的动力》都得到了很高的评价。他发现了水的"第四相"，这为新设备的创新和发明提供了巨大的机会。

图书在版编目（CIP）数据

冈特生态童书.第三辑修订版：全36册：汉英对照 /
（比）冈特·鲍利著；（哥伦）凯瑟琳娜·巴赫绘；
何家振等译.—上海：上海远东出版社，2022
书名原文：Gunter's Fables
ISBN 978-7-5476-1850-9

Ⅰ.①冈… Ⅱ.①冈… ②凯… ③何… Ⅲ.①生态环
境–环境保护–儿童读物—汉、英 Ⅳ.①X171.1-49

中国版本图书馆CIP数据核字（2022）第163904号
著作权合同登记号图字09-2022-0637号

策　　划　张　蓉
责任编辑　祁东城
封面设计　魏　来李　廉

冈特生态童书
空中的云朵
[比]冈特·鲍利　著
[哥伦]凯瑟琳娜·巴赫　绘

郭光普　译

记得要和身边的小朋友分享环保知识哦！
八喜冰淇淋祝你成为环保小使者！